T0130495

THE BENEFIT OF THE DOUBT

Deals with the universe whose true description has been long
Ignored and hidden behind the curtain of money and glory

Jaakko Kurhi

Copyright © 2021 by Jaakko Kurhi. 825283

All rights reserved. No part of this book may be
reproduced or transmitted in any form or by any means,
electronic or mechanical, including photocopying,
recording, or by any information storage and retrieval
system, without permission in writing from the
copyright owner.

To order additional copies of this book, contact:
Xlibris
844-714-8691
www.Xlibris.com
Orders@Xlibris.com

ISBN: Softcover   978-1-6641-5477-3
            Hardcover   978-1-6641-5478-0
            EBook   978-1-6641-5476-6

Library of Congress Control Number: 2021901577

Print information available on the last page.

Rev. date: 01/26/2021

CONTENTS

PROLOGUE

Welcome to my website. I must suggest that you read this entire presentation in full, perhaps twice, because my topic is unusual in nature. It could be troubling for some reader, unbelievable to other, and mind blowing for everybody else. During past years I discovered a way how the Milky Way Galaxy evolved independently of other galaxies. However, each galaxy in a realm of a universe had a similar source and similar way to begin and continue the galaxy construction. It's been forty-four years that I've thought about The Milky Way Galaxy and an atomic construction, which originate from misleading judgments by scientists who invented a particle accelerator and the Big Bang theory. Both these items lead to completely wrong direction during evaluation of an obtained result. Shame of it because true source for how the Galaxy evolved has been always available by the mechanical nature. However, it's been out of scientific observation ever since the use of telescopes.

It's not so easy. I spend more than 20 years before realizing the door to secrets of the Milky Way galaxy. My name is Jaakko Kurhi and I'm close to becoming a nonagenarian. I have nearly seven decades of working with various complicated mechanical gadgets as an inventor and visionary in the field of cinematography. With a passion for mechanics and a noted background in simplifying complex problems – my story about the invention of the revolutionary JK-103 Optical Printer is one for another day – I have been drilling into the mysteries surrounding the origin of galaxies with a unique take on the subject. If you are interested in the Milky Way theory and a new way of looking at an ancient mystery, please read on. This essay is based on how abiotic nature constructed a complex spiral galaxy, where in reality a mechanical structure requires a steady state environment to get it made. So as the story continues, pros and cons for the spiral galaxy construction come up. Hence which model is more convincing? The mathematical approach toward unknown matter, thus observing the mechanical view of possible elementary particles or perhaps simply relying on nature's way to build the galaxy and get the best results how a galaxy evolved. In order to narrow down options, the devil is in the details. Therefore, it's important to know the secrets to elementary particles that can initiate galaxy construction. The Milky Way Galaxy is unique for observation because it's the only galaxy we know well enough to get an idea of how its construction was laid out when a galaxy began to evolve. Hence I used the shape of the Milky Way galaxy and asked "How did nature builds the shape of the spiral galaxy?" The answer comes from nature's limited capacity to make something; it makes sense, during the beginning state that nature has only one type of primeval matter to work with. Thus nature has an infinite number of mass-particles to build the galaxy but it possesses nothing to build the scientifically designed atoms. So in order to understand nature's way of operating, one has to visualise what nature was at the time the galaxy began to evolve and what nature can do during the state it begins to evolve. Therefore in this approach to discover the secrets of the universe, the spiral galaxy helps to sort out fiction from reality. In other words which atomic particle's nature can naturally create and which atomic particles are results from the human act of creating subatomic particles.

So let's wind back into history where the source for the single point beginning was created and led to a misleading description of the universe. There's a popular saying in Finland: *"One who asks directions doesn't get lost"* If this phrase were put into good use more than a hundred years ago, the workings of universal matter would have been explained differently; so much so that we never would have gotten into the state where the universe began to expand from a single object. By having a hazy starting point to explain the workings of the universe, science began to work backwards in an attempt to find out which components made up the galaxy. Evidently reverse engineering with the particle accelerator and quantum mechanics didn't work well enough, because ever since the quarks and electrons found their way into atomic construction, science has used an atomic model which never had a chance of being the building blocks of the galaxy. The atom, is not only far beyond nature's building capability, but also its structural strength is too weak against the sun's extreme compression to make up the building blocks for matter. See more about this under the heading *"Atoms that make no sense"*. Comparing vital details with the results of science and nature, it begins to make sense how to approach building the spiral galaxy and its working components. It soon becomes obvious that primeval particles of matter had to be one of the kind objects and existed in clusters everywhere that observable galaxies exist presently. Whether the cluster of elementary particles was small for a single galaxy or huge one, from which several galaxies grew, the elementary particles are still of the same kind and same capability to initiate the beginning of the galaxy's construction. So how did the Milky Way galaxy begin to grow? This subject is coupled with mass particles because science finds mass everywhere in the Earth's environment, hence mass is the factor in making the galaxy also. Since mass doesn't exist without weight, and the weight is always related with the force of gravity therefore, the mass-particle has all that nature needs to create motion among other mass-particles. There's more about gravity, although it's said that gravity is the weakest of all forces in the universe. However I think this idea is linked to the prevailing atomic model which has insignificant mass thus an atom possesses no force of gravity to speak of. On the other hand, a mass-particle which nature uses is the original and solid indestructible particle, therefore the gravity is 100% stronger than the gravity of the prevailing atom. So it makes sense to use the mass particles as the beginning source for the Milky Way galaxy construction. Therefore regarding this issue I begin the galaxy construction with the primeval mass-particles (MP's) which exist in the cluster from which the spiral galaxy begins to evolve. However the most important factor is in the mechanics of how the galaxy evolves, because abiotic nature has only one way to grow, which is by chain reaction. In other words, one event leads to the next events during each state as the galaxy evolves. Another true thing is that the mechanics of the primeval mass particle cannot be more complex than what in readily is available to abiotic nature.

Let's move on: The latest detailed observation of the Milky Way galaxy is obtained from photo images, which are made available by scientific photographers and creative artists. These images indicate that the distribution of matter for the shape of the Milky Way Galaxy is due to the forces that are created during the collision-related action. Also, the photo shows an upwards-curving structure in the centre area of the galaxy. This factor itself supports evidence for the inside-out distribution of galactic matter, so obviously it takes several naturally occurring events to distribute galactic matter into an odd-shaped multidimensional spiral galaxy. Also, to make the shape of the spiral galaxy, it requires going back into a steady state environment, because moving the particle matter in the way that the spiral galaxy can evolve, isn't possible if the construction matter is in fast forward linear motion. The reason for this is that there are no forces available to change the direction and overcome the force of the fast forward motion in which the particles are moving. This is a valid argument because nature can create a working spiral galaxy in the steady state environment, but it's practically impossible to explain in detail how the galaxy evolved in a single point expanding environment. The evidence for this fact is that science informs the population about what happened in the universe, while failing to explain the details since science doesn't have any explainable details available. However, some uncertainty remains with Mass-particles also. The source for the MP's is still unknown, but it makes sense that space is not an empty place, because some sort of matter must exist before the mass-particles can exist.

So what made the mass-particles? Previously I have suggested that mass-bits exist as originator of the matter in the universe. Therefore, mass-bits remain as unexplainable phenomenon. On the other hand, science predicts the state of the absolute zero temperature must exist. Hence it makes a sense universe exist state without borders, like motionless Mass-Bit as primeval mass in the state of zero temperature, until these Mass-bits are forced by gravity into motion related activity. This action is possible because Mass-Bits possess the pull force of gravity, which is present everywhere the masses exist. So, some local regions in a realm of universe mass-bits are forced into motion which leads to accumulation action, and Mass-bits get converted into mass-particles. Therefore due these actions galaxies are everywhere in universe no matter how far we look.

Hence, let's go one step further and try to make the source of mass-bits into theoretically workable mass-particles. Years ago I spoke of Mass-Bit which size is only a fraction of the size of mass-particles. Because these mass-bits are so small they have no internal structure but they have size and weight, hence it possesses a force of gravity also. So mass-bits, each is an indestructible object which qualifies them as source for primeval matter to build up mass-particles. What's more, science speaks of two subjects which have meaningful place in very beginning state of the matter. These are subatomic particles and an absolute zero temperature of the primeval matter. Because atoms are understood as vital particles in construction of elements of the matter. Thus in this essay let's think mass-bit as a subatomic particle (SP's) which exists in the state of absolute zero temperature and as only one type of subatomic particles through out universe. So, a mass-bit and a subatomic particle represents the one and the same primeval object. Thus, this arrangement allows nature to have logically working beginning state. However, before a subatomic particle (SP's) becomes converted mass-particles (MP's), SP's has to get activated. So, SP's in zero temperature state become stimulated by pull force of gravity, which in turn accumulate SP's into small mass-particle bundles. Therefore nature has fully functional mass-particles which can further evolve into galaxy construction.

In short, nature has evolved into working elementary system, because per science, mass and force of gravity are always related, regardless how small is mass being. (Mass equal pull force). As result from above, nature is realistic being, thus a mass in any size has common attraction force. Therefore, all written above and below is within functional reality. Thus, it makes' sense, the universe exists because motionless Mass-Bits exist in the steady state environment, where mass-particles can create the event where two solid objects collide head-on and generates several factors which influence how the escaping matter gets distributed, resulting in the spiral shape of the milky way galaxy.

What's more, over a century ago the universe was considered as a steady state universe, until the Red Shift phenomenon came to light and reset the pattern of how we think about the universe. However there is more to a galaxy than the misleading Red Shift concept specifies. The fact that galaxies exist in clusters proves that the origin of matter is not from a single point distribution. It also proves evidence that universe does exist in the steady state environment, because the forceful distribution of matter from the single point source will not generate the inconsistent distribution of matter into separate clusters. Incidentally, the latest picture from the sun's surface structure indicates that similar size clumps exist throughout the sun's surface, which again suggest that nothing observable begin to evolve from the single point source. Thus, it also suggests that each sun's in the Milky Way Galaxy accumulated from multiple separate clumps of MP's, where each clump consists of zillions of MP's only. That seems to be true because atoms, molecules and elements are all multi particle objects which can't exist until the planet Earth's cooling process takes place. So, as you read you will find several items which will establish the benefits of the doubt toward the unreal universe. So the way the spiral galaxy and stars evolved, planets are born as two stars collide, atoms evolve during the cooling process of earth's crust, natural colours originate in the earthly environments, as does the red shift phenomenon. Therefore, the photon doesn't exist because light is a by-product of making an intensely hot object. Thus all the above happens because the universe exists in the steady state environment, which don't control the directional flow when matter is in motion. Also this supports the original idea of the steady state environment, to exist as place where galaxies evolve. Therefore eventually creating the event where two solid objects collide head-on and generates several factors which influence how the escaping matter gets distributed, resulting in the spiral shape of the Milky Way galaxy.

THE MILKY WAY GALAXY

The way nature built the Milky Way Galaxy in the simplest way of using only elementary particles, is an issue in this writing. Thus this is my attempt to simulate how the Milky Way Galaxy evolved in reality, which makes sense and is logically possible to be made by abiotic nature. Because the spiral galaxy is a complex system, I can't think of but one way to build it, and doing so within nature's terms. So let's build it and see how it comes out. Now that nature has been a huge cluster of primeval mass particles (MP's), and since each mass-particle in the cluster possesses a pull force of gravity, they are ready to begin the galaxy-building process. So clumps of MP's begin to grow by accumulation everywhere within the huge cluster, and because the densities of the MP's vary throughout the cluster, these clumps grow larger in denser regions. Therefore, the larger clumps begin to accumulate smaller clumps and as the gravitational force begins to gain strength and eventually causes the pull of gravity to dominate in two different regions. Hence two huge clumps with their dominating gravity remain and pull all the small clumps and any single MP's into these two growing clumps. As time passes, the huge cluster of MP's at the beginning is reduced into an empty space dominated by the gravity from the two clumps made of MP's. Consider that each MP is a solid particle, thus both huge clumps are also solid spherical objects. Finally the clumps are similar in size but located far apart from each other, yet their gravitational pull is reaching out from both clumps, thus doubling their strength of pull force (gravity) as they are moving toward each other. There's more, the rotation force of each accumulated MP's influence to the total mass in the clumps, causing them to rotate as they grow larger. Also these two clumps are highly compressed because of their heavy weight and pull force of gravity, MP's become intensely hot from the heavy weight build-up. Eventually the two huge rotating fire balls are travelling through space toward each other until they collide with destructive impact. So let's see what nature has to accomplish in order to build the Milky Way galaxy. Per photo images and scientific observations, a galaxy has two long and gradually thinning spiral arms, and the centre region is bulging up from the heavy concentration of star systems.

Also, it's said to have a large open space in the middle of the thick centre region, which is invisible behind the dense concentration of the stars. So the explained shape of the spiral galaxy gives me enough information to draw a conclusion about what has to happen during the explosively colliding distribution of the MP's, in order to come out in the shape of the spiral galaxy. After the two huge fire balls collide, the impact's effect begins to distribute the escaping MP's, and since celestial bodies rotate eastwards, this causes the axes' orientation to be from north to south. Thus it makes sense that both hot-balls of MP's rotate in a similar orientation as they penetrate into each other's colliding hot-ball. Since the impact speed is much faster than the rotation speed of hot-ball, hence most of the MP's escape in two opposite directions in 90 degree angles from the impact axes. Thus this action results in two travel paths for MP's, which in turn create two spiral arms. So the arms grow longer as hot-balls are combining into one unit and begin to rotate in the same axes speed. More happens at the time of collision. Soon after the beginning of the impact action, the impact forces gradually begin to weaken, resulting in a continuing slowdown in collision speed as the impact action continues. This slowdown in collision speed contributes to a much higher concentration of distributed MP's into the bulgy central region of the evolving galaxy. So the impact speed has slowed down but impact forces are still pushing toward the centre from two different directions, causing a back-pressure build-up in both remaining clumps, which in turn are forcing the MP's to travel backwards, leaving an open space vacuum inside that bulgy central region of distributed MP's.

Finally the MP's are distributed into the basic shape of the Milky Way galaxy, but the spiral arms are still growing in length and the bulgy region is growing in size, because the MP's are still moving outward. In order to stall this motion, the expanding speed has to slow down and finally come to a halt. Because the densest concentration of the MP's is in the bulgy region, where the force of gravity constantly increases, so then gravity begins to a slowdown, the speed of escaping MP's within the bulgy region as well the speed of two spiral arms. This action continues until the pull forces of gravity in the bulgy region begin to slow down the speed of outward-moving MP's, until the force of gravity and the force of expanding motion come to an equilibrium state space. In this state all expanding motions have stalled and the Milky Way galaxy has reached its size and shape. So the time passed by and nature has distributed MP's everywhere within the Milky Way cluster, which is now much smaller than the one from which the Milky Way Galaxy began to evolve. Thus gravity is the only force nature operates with, and has used it to create motion to make two huge spheres of MP's, then it explosively distributed MP's into the spiral-shaped galaxy cluster, from which nature accumulate smaller clusters of MP's, hence nature uses gravity and motion again to accumulate MP's into stars within the evolving spiral galaxy. One may wonder why stars are made from MP's only. There are two reasons for this, It's because the atoms, molecules, and elements of matter are combined particles; hence anything but MP's will be destroyed by the star's extreme compression force and high temperature. Thus atoms and related particles don't exist until planets are cooled to the state where the temperature is lower and the force of extreme compression is lower than the atomic bonding force of gravity. Another reason is, because in this theory an atom is a bundle of mass-particles, therefore an atom is a destructible particle in extreme conditions in the galaxy. On the other hand MP's are permanent elementary particles which can't be destroyed in any condition. So the spiral galaxy is in the state where stars begin to grow, and because the MP's possess the pull force of gravity, they begin to accumulate into a small clump throughout the spiral-shaped cluster. In this state the combined force of gravity has more strength and begins to pull clumps together into large clumps. This cycle continues until all clumps and single mass-particles are accumulated into a spherical-shaped hot and bright object we call Stars.

Also as each star grew larger and heavier, they became intensely hot. Thus, the Milky Way shaped cluster of MP's, became the host for 400 billion hot and visible stars. So, nature has stars throughout the galaxy but planet earth is still missing. So, how to break a star in liquid state, so that several small portions of the star breaks off into hot clumps, which then became the planets in the solar system. Nature built the spiral shaped galaxy, using collision force for forming its spiral shape. Also, stars accumulated in different size throughout the galaxy by the force of gravity. So, nature will use colliding force and the pull-force of gravity again, for assembling the solar system. Because stars in the galaxy are intensely hot, they are in the liquid state. Thus, two nearby stars by their combined gravity, began moving toward each other and eventually came into a softly colliding action. This action is possible because stars were so close to each other, they had no time to accelerate into high speed travel. So, two colliding liquid balls can softly brake off several small portions of liquid clumps from both colliding balls into outward motion. During this colliding action remaining portions of each colliding star blended into a single larger spherical star which in this case become our Sun. Hence we have a stable rotating Sun and outwards moving hot planets. Thus, all planets are moving outwards in different rotation and linear speed. The way liquid clumps brake off from the each colliding star during unevenly applying collision forces. So the gravity from the Sun and each planet, force outward motion to stall with varying force thus hot planets' finds their orbiting position around the Sun indifferent distances from the Sun. So this colliding action explains the source for planets and their moons, and an asteroid belt among the planets. Finally we can concentrate on what can happen in Earth's environment; since the sun is made from MP's only, then it makes sense that all planets orbiting the sun is made from MP's also. Because atoms are destructible particles, under extreme compression; hence they can accumulate only near the surface region of the Earth's crust, which is still in the liquid state but the temperature and compression force is much lower. A long time elapsed as Earth cooled and the crust began to build up. In the meantime atoms have evolved by accumulation into bundles of MP's. These bundles come in various numbers of MP's, hence different sized atoms combine into molecules for all minerals, gases, and chemical elements. Therefore, before the crust becomes the solid matter, light molecules escape above the crust, forming the earth atmosphere. So in the Earth's crust and everything above crust are made of the molecules in the different element values. On the other hand, the hot and heavy inner core of the planet earth remains as a huge ball of mass-particles in liquid state. Therefore it makes sense, suns and planets are made of mass-particles and atoms exist basically in molecule format in the planets' crust and above a crust region only. So, nature has proved that an atom is not a complex systems, because simple primeval mass-particles can build the Milky Way Galaxy and its stars with planets where atoms and molecules exist.

The way nature built the Milky Way Galaxy in the simplest way of using only elementary particles, is an issue in this writing. Thus this is my attempt to simulate how the Milky Way Galaxy evolved in reality, which makes sense and is logically possible to be made by abiotic nature. Because the spiral galaxy is a complex system, I can't think of but one way to build it, and doing so within nature's terms. So let's build it and see how it comes out. Now that nature has been a huge cluster of primeval mass particles (MP's), and since each mass-particle in the cluster possesses a pull force of gravity, they are ready to begin the galaxy-building process. So clumps of MP's begin to grow by accumulation everywhere within the huge cluster, and because the densities of the MP's vary throughout the cluster, these clumps grow larger in denser regions. Therefore, the larger clumps begin to accumulate smaller clumps and as the gravitational force begins to gain strength and eventually causes the pull of gravity to dominate in two different regions. Hence two huge clumps with their dominating gravity remain and pull all the small clumps and any single MP's into these two growing clumps. As time passes, the huge cluster of MP's at the beginning is reduced into an empty space dominated by the gravity from the two clumps made of MP's. Consider that each MP is a solid particle, thus both huge clumps are also solid spherical objects. Finally the clumps are similar in size but located far apart from each other, yet their gravitational pull is reaching out from both clumps, thus doubling their strength of pull force (gravity) as they are moving toward each other. There's more, the rotation force of each accumulated MP's influence to the total mass in the clumps, causing them to rotate as they grow larger. Also these two clumps are highly compressed because of their heavy weight and pull force of gravity, MP's become intensely hot from the heavy weight build-up. Eventually the two huge rotating fire balls are travelling through space toward each other until they collide with destructive impact. So let's see what nature has to accomplish in order to build the Milky Way galaxy. Per photo images and scientific observations, a galaxy has two long and gradually thinning spiral arms, and the centre region is bulging up from the heavy concentration of star systems.

Also, it's said to have a large open space in the middle of the thick centre region, which is invisible behind the dense concentration of the stars. So the explained shape of the spiral galaxy gives me enough information to draw a conclusion about what has to happen during the explosively colliding distribution of the MP's, in order to come out in the shape of the spiral galaxy. After the two huge fire balls collide, the impact's effect begins to distribute the escaping MP's, and since celestial bodies rotate eastwards, this causes the axes' orientation to be from north to south. Thus it makes sense that both hot-balls of MP's rotate in a similar orientation as they penetrate into each other's colliding hot-ball. Since the impact speed is much faster than the rotation speed of hot-ball, hence most of the MP's escape in two opposite directions in 90 degree angles from the impact axes. Thus this action results in two travel paths for MP's, which in turn create two spiral arms. So the arms grow longer as hot-balls are combining into one unit and begin to rotate in the same axes speed. More happens at the time of collision. Soon after the beginning of the impact action, the impact forces gradually begin to weaken, resulting in a continuing slowdown in collision speed as the impact action continues. This slowdown in collision speed contributes to a much higher concentration of distributed MP's into the bulgy central region of the evolving galaxy. So the impact speed has slowed down but impact forces are still pushing toward the centre from two different directions, causing a back-pressure build-up in both remaining clumps, which in turn are forcing the MP's to travel backwards, leaving an open space vacuum inside that bulgy central region of distributed MP's.

Finally the MP's are distributed into the basic shape of the Milky Way galaxy, but the spiral arms are still growing in length and the bulgy region is growing in size, because the MP's are still moving outward. In order to stall this motion, the expanding speed has to slow down and finally come to a halt. Because the densest concentration of the MP's is in the bulgy region, where the force of gravity constantly increases, so then gravity begins to a slowdown, the speed of escaping MP's within the bulgy region as well the speed of two spiral arms. This action continues until the pull forces of gravity in the bulgy region begin to slow down the speed of outward-moving MP's, until the force of gravity and the force of expanding motion come to an equilibrium state space. In this state all expanding motions have stalled and the Milky Way galaxy has reached its size and shape. So the time passed by and nature has distributed MP's everywhere within the Milky Way cluster, which is now much smaller than the one from which the Milky Way Galaxy began to evolve. Thus gravity is the only force nature operates with, and has used it to create motion to make two huge spheres of MP's, then it explosively distributed MP's into the spiral-shaped galaxy cluster, from which nature accumulate smaller clusters of MP's, hence nature uses gravity and motion again to accumulate MP's into stars within the evolving spiral galaxy. One may wonder why stars are made from MP's only. There are two reasons for this, it's because the atoms, molecules, and elements of matter are combined particles; hence anything but MP's will be destroyed by the star's extreme compression force and high temperature. Thus atoms and related particles don't exist until planets are cooled to the state where the temperature is lower and the force of extreme compression is lower than the atomic bonding force of gravity. Another reason is, because in this theory an atom is a bundle of mass-particles, therefore an atom is a destructible particle in extreme conditions in the galaxy. On the other hand MP's are permanent elementary particles which can't be destroyed in any condition. So the spiral galaxy is in the state where stars begin to grow, and because the MP's possess the pull force of gravity, they begin to accumulate into a small clump throughout the spiral-shaped cluster. In this state the combined force of gravity has more strength and begins to pull clumps together into large clumps. This cycle continues until all clumps and single mass-particles are accumulated into a spherical-shaped hot and bright object we call Stars.

Also as each star grew larger and heavier, they became intensely hot. Thus, the Milky Way shaped cluster of MP's, became the host for 400 billion hot and visible stars. So, nature has stars throughout the galaxy but planet earth is still missing. So, how to break a star in liquid state, so that several small portions of the star breaks off into hot clumps, which then became the planets in the solar system. Nature built the spiral shaped galaxy, using collision force for forming its spiral shape. Also, stars accumulated in different size throughout the galaxy by the force of gravity. So, nature will use colliding force and the pull-force of gravity again, for assembling the solar system. Because stars in the galaxy are intensely hot, they are in the liquid state. Thus, two nearby stars by their combined gravity, began moving toward each other and eventually came into a softly colliding action. This action is possible because stars were so close to each other, they had no time to accelerate into high speed travel. So, two colliding liquid balls can softly brake off several small portions of liquid clumps from both colliding balls into outward motion. During this colliding action remaining portions of each colliding star blended into a single larger spherical star which in this case become our Sun. Hence we have a stable rotating Sun and outwards moving hot planets. Thus, all planets are moving outwards in different rotation and linear speed. The way liquid clumps brake off from the each colliding star during unevenly applying collision forces. So the gravity from the Sun and each planet, force outward motion to stall with varying force thus hot planets' finds their orbiting position around the Sun indifferent distances from the Sun. So this colliding action explains the source for planets and their moons, and an asteroid belt among the planets. Finally we can concentrate on what can happen in Earth's environment; since the sun is made from MP's only, then it makes sense that all planets orbiting the sun is made from MP's also. Because atoms are destructible particles, under extreme compression; hence they can accumulate only near the surface region of the Earth's crust, which is still in the liquid state but the temperature and compression force is much lower. A long time elapsed as Earth cooled and the crust began to build up. In the meantime atoms have evolved by accumulation into bundles of MP's. These bundles come in various numbers of MP's, hence different sized atoms combine into molecules for all minerals, gases, and chemical elements. Therefore, before the crust becomes the solid matter, light molecules escape above the crust, forming the earth atmosphere. So in the Earth's crust and everything above crust are made of the molecules in the different element values. On the other hand, the hot and heavy inner core of the planet earth remains as a huge ball of mass-particles in liquid state. Therefore it makes sense, suns and planets are made of mass-particles and atoms exist basically in molecule format in the planets' crust and above a crust region only. So, nature has proved that an atom is not a complex systems, because simple primeval mass-particles can build the Milky Way Galaxy and its stars with planets where atoms and molecules exist.

ATOMS THAT MAKE NO SENSE

Abiotic nature is about the lack of life, meaning that everything that happens during galaxy construction occurs in mechanical order. Therefore nothing evolves by accident or for good reason. Instead abiotic nature operates like a chain reaction, where by one event causes the next events to happen automatically. Hence, nature-made atomic construction is limited to primeval mass-particles only. Thus as a galaxy evolves, nature uses mass-particles instead of atoms as the building blocks in every phase the galaxy evolves. So let's compare nature-made simple building blocks to atoms, which are assembled as per obtained information on the particle accelerator and quantum mechanics. Hence this experiment leads into an atomic construction, whose complexity science can't explain in regard to how atoms evolved. For example how can one explain Quarks which evolved one terasecond after the Big Bang, at the same time evolving into six different types of quarks which have different energy values. That's all - no clarifying details of how quarks and their complex system evolved, yet the complete atomic system is based on quarks.

|  | Scientific atom is basically empty open structure, has no mechanical support for nuclei and circling electrons | The image of a solid heavy atom makes sense because matter needs solid atoms to be the building blocks for matter. | 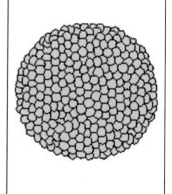 |

This heavy atom is made of solid unbreakable mass particles which can accumulate into elements of the matter during earth's cooling process. So what can be said to defend the scientific atom which is not a practically workable particle. There is another questionable item regarding electrons in the atom, because it makes no sense that they don't have any structural strength to speak of. The orbiting electrons in their shells have insignificant mass, so they lack supporting mechanical structure, yet the presence of orbiting electrons accounts for more than 99% of the total empty atomic space. Also the electron cells have no mechanical structure, nor does the nucleus give any structural support, because it's sort of floating in the open space in the middle of circling electrons.

So, in a mechanical sense, the insignificant mass of five circling electrons separate each atom, everywhere the matter exists. What's more, in order to get a visual idea of the atoms' emptiness, consider this: if a nucleus is a tiny pencil dot, then the electrons are circling some three long steps away from the nucleus. So it's not possible for such empty atoms to stay together without getting crushed inside Earth's molten core. Because planets' core is under extremely high pressure, it generates Earth's internal temperature near 5000 C. Hence, a scientific atom has no chance to exist in environments of Sun and planet Earth. In fact atoms evolve when planet earth' crusts has cooled down until a thick surface layer allows mass particles began to accumulate into various size atoms. As expressed before, the mass particle is a solid object therefore cannot be destroyed by any means, and because The Milky Way Galaxy evolved strictly from the primeval mass particles, thus all Suns we observe are made of mass particles only. Thus in conclusion, these lightweight atoms don't seem like a good idea, and most importantly abiotic nature has no resources to make such atoms. What one sited science have found as useful information doesn't work well when comparison is drawn in between science and nature? After all nature operates mechanically hence its predictable what mechanical nature can do and cannot do as it evolves. Therefore, re-evaluating atomic construction would advance the scientific view of atoms to the level at which nature operates.

IS A PHOTON REAL OR NOT

This article is based on the benefit of the doubt, generated from scientifically explained state of light. What is light? It's still a lingering subject, so I use nature to explain how light originates. Let's begin with nature's limited capabilities, because all of nature's operations are based on simple mechanics, having no resources to build or design anything. Therefore all nature can do, is to carry through action which leads to a following action that is naturally available. Thus, simplicity prevails in all naturally-evolving actions. So let's look at light from nature's point of view and eliminate what light is not. Light has no mass, so it is not a particle or ray. For the same reason, light is not a unit or an existing entity. Light can't be permanently stored or maintained in any form of matter. Also, light doesn't possess the source of electromagnetic radiation, because lacking mass, electromagnetism can't exist. Finally, light is not a photon because nature can't make multi-functional packet-type particles, not by design or accidental evolvement. Hence, a photon is not a useful particle because colours originate in the Earth's environment, instead of coming from sun in the packet format. So due to the fact that the above-mentioned items don't work with nature, let's try a different approach and follow the trail left by nature. Everywhere we observe light, it is in an illuminating format. In other words, light is in a transparent state when it travels between the source and the object.

However when light lands on the object, it illuminates this object into an observable state. So whether it's a landscape or a room full of objects, they all are illuminated by invisible light when it lands on them. In short, light is transparent like glass, while in transit between the source and these objects. This phenomenon explains why we don't see light travelling between the source and an object nor do we see the reflected light travelling between illuminated objects. So what is light? Let's begin with sunlight which nature made by accumulating together so many mass-particles that the core of the growing ball began to heat up due to its enormous weight, and eventually became so heavy and hot that it appears like a fire ball in the sky. So nature made an intensely hot sun which illuminates all the surrounding planets. Comparatively sources such as incandescent light bulbs, candlelight, florescent lights, burning flames and LED light, all work the same way as the sunlight, having an intensely hot source, bright enough to illuminate surrounding objects. So it seems that light is not a readily available unit, instead the sources for light have to be produced or exist before light appears as an illuminator, regardless of what type the light source is. Hence light can't be a basic unit or elementary particle which exists everywhere light is required. Therefore the source for light is either naturally made like the sun, or a man-made bright source which is turned on and off at will. In fact one can't turn the light off unless turning the bright source off. So in reality light doesn't need to be stored as an elementary particle or possess electromagnetic radiation which has no functional purpose. I always wonder when the candle light is lit where does the electromagnetic radiation come from?

So adding up all of the above observations, an intensely hot spot has to be made before invisible light becomes an illuminator. For this reason there is no other logical conclusion but that light is a "by-product" of making these shiny hot objects. So light is merely an invisible flow of reflection from the intensely bright surface which becomes visible when it lands on a tangible surface and illuminates this surface. There is another misconception in dealing with light, as demonstrated by candlelight. It's an obvious fact that the colour spectrum is an earthly creation because the flame of the candlelight begins clear like glass without colour, the wick and the background are clearly visible through the flame, which quickly turns into blue-yellow-orange-red as molecules from the air enter into the candle flame. This fact is true because the only source for these colours is in the Earth's atmosphere which is saturated with an invisible mixture of more than seven gases existing in the air, and make up the colour spectrum. This factor confirms that natural colours originate from the Earth's environment. Can light be this simple? I think it can because the by-product cannot be an elementary particle, and electromagnetism requires a constant source to electricity related energy. Also all useful features of light including natural colour are enhanced within the Earth's environment, where invisible light goes through various natural colour filters. Therefore, these factors make photons useless particles, whose format in reality nature cannot generate or assemble. Hence the benefit of the doubt is enhancing traditional thinking towards nature's capacity to operate.

RED SHIFT PHENOMENON

Let's go little further in evaluating process how light and colours are related, now that in this essay light is observed as by-product, which in manufacturing process often is discarded as disposable throw away. However, a nature has some good uses for this light and because the source for light is fairly easy to manufacture, lot of earth's surface is illuminated in various degree during dark hours. Although our sun is the most efficient natural source for light, yet all light sources behave the same way as sunlight. So light is continuously flowing bright but invisible streams in between source and illuminated object. Yes it's puzzling but true, because during light travels in between source and object it's in transparent states, but when it lands on an object, light becomes an illuminator. In other words, light goes by our eyes at the speed of light all the time, but we don't see it. Thus, light doesn't travel as a wave or any kind matter. What's more, light have no mass particles, therefore light doesn't have size and weight. Here is another important factor to consider when evaluating the Red shift phenomenon. The colours spectrum which science applies into red shift theory has limitations because all colours formats has mass involved. In other words, all colours originate within mass particles. So, light is not the source for colours instead matter creates colour. However, the question is what make colour? The earth's surface and atmosphere are good colour's producers but sunlight light we receive, make no colours, in fact in various decrees colours are property from planets circling the sun, especially our well-known planet earth is bursting with colours.

Consider this, within planet earth environment, natural lights are colourless at the beginning from the source, whether sources are sunlight or man-made. Therefore as natural light travels through space in between source and object, it carries no particles or any properties other than it illuminates all objects it lands on. In other words, light is invisible to an observer during it travels through space in between source and object whose colours it's illuminating. Hence, there is no colour spectrum available on the sun's surface nor from the distant galaxy. Therefore, natural light doesn't add colours. Instead it illuminates coloured surfaces. Obviously this old observation isn't included in the Red Shift theory, because the theory is based on red and blue colours indicating whether a distant galaxy is moving away or toward the observer. However, based on all above, there are no colours nor colour waves to observe and get support for red shift theory. So how this works out, the light beam from the distant galaxy originates from a huge area of stars, which are reduced into an observable tiny hot and bright spot. So, from this bright spot only mass less light is observable. So, that's how light behaves regardless if light comes from natural or a man-made source. Hence viewing from the most workable point of view, the universe operates in the steady state format, which aspect works well with the mechanically operating nature. For a conclusion, light from the sun or the distant galaxy don't include colours, until this light beam illuminates the colours object. In other words, for colours to appear natural light requires to land on a colour surface which always happens in earth's environment. So from the distant galaxy in the deep space, light of the zillion stars is combined into a tiny spot and photographed in B&W film. Because red or blue colours can't be observed on B&W film. Hens, changes are colours are generated during post production of the B&W image. However, it's puzzling, because it's not well expressed how the colour-spectrum of light-waves is observed because in reality, light is mass-less as it propagates from source to matter.

SUPPORTING EVENTS FOR A SPIRAL GALAXY

Ever since people developed ways to express their thoughts the mystery of the Heavens has been the constant subject of conversation. Currently perhaps three million years later, modern-day people are using advanced observation equipment, yet the same view of the sky remains an unsolved mystery, from which all knowledge of visual and invisible matter has to be extracted from. However in my opinion I have demonstrated that the Milky Way Galaxy can be explained in a logical format. Also the way the spiral galaxy evolves in this story, it's mechanically doable during each event that happens automatically, as chain reactions continue until the complex spiral galaxy has matured into observable state. So the benefit of the doubt lingers over the scientific description of universal matter. The scientific model of an atom is notable for its complex construction for two reasons. In the first reason science explains subatomic particles and their function in precise details, but how these subatomic particles come to exist is beyond any attempt to explain their origin. The second reason which has more disturbing problems comes from electrons which have no place in an atomic construction as explained in the Milky Way text. So, all universal objects whose core and crust are under extreme pressure, can't possess particles which are made of subatomic particles. Therefore scientifically designed atoms which are like a huge empty cell, but have an insignificant nucleus, hence during beginning state, these complex atoms can't exist before the planets are in the cooling state.

So, if a beam of light and quantum mechanics don't deliver what we are looking for, the answers should be found elsewhere in a simplified format. Obviously, I am talking about numerous alternate versions written against the popular Big Bang model, and all for the same reason, the working mechanics of the Big Bang model are not convincingly explained, thus creating follow-up questions to which there are no mechanically workable answers. So let's dig deeper into the beginning of matter and sort out a workable source which is unknown but must exist before MP's can exist. Also because light is the by-product of these bright and hot celestial objects, there is no obtainable information available, which has influence on how each galaxy operates. Per science the Higgs boson may have been identified, so I use its size to compare, what is the smallest existing thing that makes solid mass-particles which when combined into clumps, become atoms in the planets of the solar systems. So it's safe to use Mass-particles as the main factor, in every state as the galaxy evolves. Keep in mind, atoms made of mass-particles MP's, as well as atoms having circling electrons cannot exist as individual atomic objects under extreme pressure. For example, stars are made of MP's because each star is always in the state where the extreme temperature and high compression forces destroy all structural construction. On the other hand, the planet's environment is more suitable for MP's to accumulate into atomic bundles of various size. So in conclusion, nature made MP's which built the spiral galaxy, while the scientific atom had no chance to go through the first important evolving action where the MP's were distributed into the shape of the spiral galaxy. It's difficult to accept an idea where something began to evolve from an empty space vacuum. However, because the sources for matter exist, which in this case are established as MP's therefore these MP's have to have a source from which they are created. Now we have the legitimate setup to begin the galaxy construction. Like a carpenter said, I don't have to know where the lumber comes from but I can still build the house from this lumber. So nature has, MP's which will carry on during each evolving operation as galaxy grows.

What is atomic energy? Science among other subatomic particles found low and high energy in each atom; also atoms possess gluon which is energy that keeps the atomic system together. On the other hand nature' atomic energy is the gravity, because each MP is made 100% of mass and mass equals pull force of gravity, therefore the pull force of gravity increases as number of MP's increase in the atomic bundle. So heavy atoms are still a valid concept for atomic bombs, as well as lighter atoms work for all elements of matter. Thus, the number of MP's per atomic bundles controls the atomic weight of each element in the periodic table. So nothing is lost with the mass-particle theory, but lots of misleading abstracts. The spiral galaxies in a realm of the universe are basically copies of repeated initial actions occurring as each galaxy begins to evolve. That's because the beginning source is always a huge cluster of MP's, regardless of the clusters' independent position in natures' illusive space. In other words, space which is considered an empty vacuum, is not empty and it's not a vacuum either, because a vacuum is associated with the enclosure which is made empty in earthly terms. Thus, space isn't an enclosure, instead it's something which is busting up our brain in the quest for better answers. Nevertheless, the Milky Way's originate from the real mass which has to have a beginning source. So, it can be said that the atomic-like objects that may look similar to a ball of sand, are basically clumps of mass-particles, which accumulate in various numbers of these mass-particles.

Some words about the invisible void inside the bulgy region of expanding MP's: If you wish to observe the explosively driven action during the distribution of MP's into the shape of a spiral galaxy, simply hold a rubber band with both hands. Your left hand moves toward the left in steady speed while the right hand should follow in the same direction, both simultaneously, but the speed of the right hand gradually slows down. The area left behind your right hand represents a void, while the band's tip which you will be holding by your left hand would represent the expanding and fastest-moving matter. Incidentally, this action cannot coincide with the description that universal matter was distributed from one small object or from a collection of objects from a small original source. Finally I come to the most confirming factor, concerning how the galaxy evolves. Matter in the steady state environment is more adaptable for mechanically evolving changes, vs. linearly moving matter, which has no forces available to overcome the energy of the motion, and to change the direction of extremely fast moving particles of matter. So I have demonstrated how galaxies throughout the universe can evolve in a mechanically logical way. However there are obstacles to overcome, like my scientifically educated friend said: Even if you are right in the galaxy concept they will not change the current theory in a million years. On the other hand, modern people are getting smarter by leaps and bounds, so the next generation of text book readers may be protesting and demanding more detailed and workable information about the state of The Milky Way Galaxy. Problems with atoms, light is a by-product, a photon which nature cannot make, And the most misleading concept, the Big Bang theory. So, next generation of space scientists have some make up work to do.

ABOUT THE AUTHOR

Jaakko Kurhi is a Karelian born Finn that possess a lot of experience in complex mechanics. Professionally, Kurhi is an optical instrument designer and builder for which independent film makers are happy round the world. His hobby is to tangle with interesting and meaningful subjects, especially items where mechanics are a major factor in the systems operation.

Printed in the United States
By Bookmasters